Beating Your Competition
Through Quality

Beating Your Competition Through Quality

D. B. Owen
Southern Methodist University
Dallas, Texas

Marcel Dekker, Inc.   New York and Basel

To my wife

ISBN 0-8247-8065-5

This book is printed on acid-free paper.

MARCEL DEKKER, INC.
270 Madison Avenue, New York, New York 10016

Current printing (last digit):
10 9 8 7 6 5 4 3 2 1

PRINTED IN THE UNITED STATES OF AMERICA

Preface

THIS BOOK IS WRITTEN FOR THE INDIVIDUAL WHO HAS LITTLE OR NO feeling for what should be done when variation is encountered. It is not intended for the quality control expert nor for the statistician. It is for the people who are just becoming aware that superior quality leads to great competitive advantage. This book tells how superior quality is attained and gives some of the basic techniques involved. It gives clear insight into the impact of variation on the future of the world.

Most things we do and most things we make vary. In some disciplines an attempt is made to eliminate variation without giving any thought to the cost involved. In other cases it is impossible to eliminate the variation, and in many of these cases people are baffled by how to handle the situation. In the United States little is taught in the schools to help people with this problem.

Japan's emergence as a major manufacturing power now requires that people understand variation if we hope to compete on an equal footing. Of course, Japan's strength has been attributed to many factors. Among them are: (*a*) robots; (*b*) government support of industry through investment incentives and savings incentives for workers; (*c*) greatly increasing levels of education for its populace; (*d*) lifetime employment in many industries, which leads to a willingness for workers to accept automation; (*e*) "quality circles," which involve feedback from workers for better production procedures, i.e., the worker is made a part of the management team; (*f*) office automation as well as factory automation; and (*g*) quality control through statistical methods, also known as statistical process control (**SPC**).

Each of these factors contributes to Japan's success, but none of them is more overriding than quality control through statistical methods. Government support of industry and lifetime employment existed in Japan in the 1940s, when Japan was noted for turning out shoddy goods. The United States has more robots than Japan, but there is some question as to the sophistication involved. The United States is well along in office automation and may be ahead in this area, but the United States has failed in quality circles partly because the American worker does not understand variation. Our educational system has failed to educate more than a few in the ideas of how to deal with product variation, how to minimize this variation problem, and the statistical methods associated with it.

Variation is more widely understood in Japan because statistical science is taught at an early age. The Japanese do not fully realize that their main strength has come from this area. Most Japanese believe that the United States must be as far along as Japan because the ideas for eliminating variation were developed in the United States. Two people—both consultants—are given the most credit by the Japanese: W. Edwards Deming, from Washington, D.C., and J. M. Juran, from Connecticut. The Japanese have even named a prize after Deming, given annually to the company using the best statistical methods. This prize is highly coveted.

It is my hope that this book will open the eyes of people in the United States and other countries to the problems of variation and provide some information on very elementary solutions to problems in variation. The book is not intended to serve as a handbook of solutions. It is for the layman, but it cannot make anyone an expert. I hope it will cause many chief executive officers of companies to reevaluate how they are handling their data.

Is all of the information in your data being extracted and put to use to solve your problems? Probably not, and this could eventually lead to the loss of your job. Since manufacturers have been the principal users of quality control techniques, many people think of these techniques as being a manufacturing tool only. That is absolutely wrong! All jobs are affected, including the service industries and the professions. For example, these methods are being used more and more in courts of law in the United States.

One other problem that occurs in this field is that there are many people who claim expertise in statistics but are only partly prepared to handle the many problems. I am talking about the statistical problems, not the various engineering and management problems that occur in producing goods. The manager and the engineer are generally not prepared to handle the problems in variation in any but superficial ways. (See more on this issue in Chapter 15.)

The importance of addressing statistical problems constitutes the focus of the book. Many different disciplines are used to produce a quality product, but the one element that has set Japan apart from the United States is the general understanding by the Japanese of statistical methods. Engineering and management are separate issues, although

Preface

there are obvious interactions with the statistical methods. I will comment on some of these interactions in the text.

I welcome any comments. I hope the readers of this book enjoy it.

D. B. Owen

Contents

Preface

1
Introduction

MANY COMPANIES, AND WHOLE INDUSTRIES, ARE FIGHTING FOR THEIR continued existence in the face of overwhelming competition, that often provides better quality than these companies are able to produce. These managers want to know how to achieve this better quality immediately. Unfortunately, the achievement of outstanding quality cannot occur overnight. It took the Japanese more than 20 years to reach their present preeminent position. So, what is it that the Japanese know that United States companies do not know? The Japanese took to heart the coaching of W. Edwards Deming, an American statistical scientist who went to Japan after World War II. The U.S. companies paid only token attention to what Deming and his colleagues were saying and are now paying the price for their inattention. The Japanese themselves also are being challenged by the nations of Korea, Taiwan, and others that have been watching and listening to what the Japanese have done. The rest of the world, including the United States and Canada, must follow suit or they will be forced to provide a lower standard of living for their people.

Chapter 1

In recognition of the knowledge imparted to them by W. Edwards Deming, Japanese companies give the annual Deming Award to one of their industries that has achieved the most significant improvement in quality. Before World War II the label "Made in Japan" was often used as a synonym for poor quality. In the 1980s "Made in Japan" has become a symbol of quality. What is it that Deming and the other consultants taught the Japanese, and which the rest of the world wants to learn?

The fundamental idea that pervades all of this is the concept of variation and how to deal with it. First, we have to realize that everything that we make or do varies. Sometimes we have to have delicate measuring instruments to detect the variation, but it is always there. But, you say, didn't they teach us in school that the speed of light, for instance, is constant? The speed of light may be constant, but our measuring instruments have given us differing values for the speed of light as the measurements have been refined. When we finally refine our measurements sufficiently, we may find that there is a small variation in the speed of light. The point is that we should be prepared to deal with variation, and generally speaking, we have not been trained to do so. More will be said about this in Chapter 4.

Throughout this book the language will be that of the manufacturing industries, because that is where most of the action has occurred. In fact, the techniques involved apply to all businesses, including the service industries. The companies that adopt these techniques first in the service industries will gain an almost insurmountable advantage over those that do not, just as the Japanese electronics manufacturers have done compared to the U.S. companies. The minimum time to bring enough of the people in a company on board to see some changes in quality with the new techniques is at least two years and may be as long as 20 years.

Introduction

Many companies will never catch up if they let the initiative in this area go to their competitors. Remember, it is over 30 years since the Japanese were first introduced to these techniques, and they have been refining them ever since. It is not a simple matter to apply the technique to a different company or industry, but it can be done and it will be done by those who will become the leaders in the next century.

In the United States, Canada, and many other nations of the world, alarms are being sounded and managers are being forced to make adjustments to the new type of competition. In the United States we see many instances of one-quarter to one-half of the work force devoting their time exclusively to fixing mistakes that have been made. Included in these percentages are the managers who spend most of their time trying to placate irate customers. The need to change so that we make the product right in the first place is evident for many American products. Many products need to be redesigned, statistical process control techniques installed, and workers given a measure of control over their lives and a sense of pride in what they do.

Another key element is that we must first determine whether the variability in a process is in control. The limits of control should be set by the process itself. If the process is not in control, then we must find the assignable causes for the lack of control and eliminate them. Once we have control, we compare the output of the process with the specifications (if any) or the goals we wish to attain and decide whether the process needs further tightening up. It is important to understand a fundamental point here. If the process is not in control, there is some possibility that the worker could be the culprit. However, if the process is in control, then management must do something to tighten the process up.

Even if all the specifications are met, it may be desirable to try to eliminate some of the causes of variation. There is the story of the transmissions for Ford cars: one lot was made in the United States and another lot was made in Japan. Both sets of transmissions met the specifications. They were tested over many months and the Japanese-made transmissions had much better repair records than those made in the United States. At the end of the test the transmissions were torn apart and carefully measured. The conclusion was that the Japanese-made transmissions had been manufactured much closer to target values than those made in the United States. The lesson to be learned is that just meeting specifications is not enough. We must constantly work toward reducing variability. There is also a rule of thumb that experienced quality control people often quote. Approximately 80% of the variation on most products may be traced to four to six causes, and the remaining 20% of variation may be due to many, many small causes. It is important to concentrate effort on the four to six important causes first and not spend critical time on the many trivial causes of variation. The biggest payoff comes with the few.

So how do you find out what the causes of variation are and which ones are contributing the most to the variation in the product? We will attempt to answer these questions in a general way in this book. Specific situations often require unique answers to these questions.

Since the mid-seventies many people have traveled to Japan and thought they found the reason for Japan's resurgence after World War II, because there are many differences between the cultures of Japan and the United States. Among the differences noted were the cost of labor, the tax structure, the availability of capital from savings by the population, and so forth. However, what was overlooked and is still being overlooked is that historically most of these differences existed in varying degrees before World War II, when Japan's quality was poor.

In the late 1980s a sweeping overhaul of management at the corporate level is underway. To keep up, top management must maintain a constant push for quality. If the push is not maintained by top management, the workers conclude that any effort is a sham. Also, the group that is most resistant to the quality push is middle managers. They have the feeling that it takes away from their authority. In fact, it changes the role they play by requiring that they spend more time in planning the work to be done and less time bossing people. Providing the appearance of quality is easy, but actually attaining quality is a long-term, difficult job. The cost of attaining quality may seem high at first, but the return on investment is greater than almost any other return a business can realize.

The people at the working level are usually eager to assume the responsibility for the quality of the items they produce. In many industries, inspectors are being eliminated, and the responsibility for inspecting the product has been given to the worker producing the product. Of course, there must be monetary recognition for the added work, but it has generally been found that total costs have declined with this system.

There has traditionally been a wall between the design engineer and the manufacturing engineer in many industries. Production of quality goods requires that the entire process—from design through manufacturing to field service—be treated as a single entity: i.e., a single integrated plan must be developed instead of separate plans for each of the stages. The goal is to satisfy the customer. Every stage has to consider the customer and what he expects from the product. This procedure of treating the whole system and not just the segments is called statistical process control, sometimes referred to as SPC. The word *statistical* comes into play because only sophisticated statistical analyses can sort out the maze of relationships involved.

New technology, including robotics, can have a big effect in the race for quality, but new technology is not the most important component. More important than technology is changing the management structure to insure that quality products be made. Quality improvement must be a formal part of the business plan. In many companies the annual performance review now includes a review by the director of quality for the company who certifies the extent to which the person involved practiced good quality in his job. More will be said about this aspect in Chapter 10.

2
The History of Competition from Japan

BEFORE WORLD WAR II, JAPAN WAS NOTED FOR PRODUCING SHODDY goods, and Japanese firms tried several subterfuges to keep the buyers of their goods from realizing their products were made in Japan. One Japanese company built a factory in a city named Usa and prominently marked its goods "MADE IN USA." At that time the United States clearly led the world in quality of goods. It is generally not true that U.S. quality became poorer, but, in fact, quality from Japan became much, much better than it was.

How did this come about? Several consultants went to Japan from the United States right after World War II to help the country recover from the war. Among them were W. Edwards Deming and J. M. Juran. Both Deming and Juran are still conducting seminars (at the time this is being written, in June 1988) on what they taught the Japanese and what they learned from the Japanese in turn. The message taken to Japan was simply, "You must build quality goods, and these are the techniques that the United States has developed for doing this." The techniques outlined are those of statistical science.

The field of statistical science has grown tremendously since the end of World War II, with many new techniques now being available. There is now (in June 1988) an encyclopedia with eight volumes already published, and more to come, devoted exclusively to short articles on statistical science, which at this writing contains over 5000 large printed pages. At the end of World War II only a few books were available to students of statistics. Most of this development has been in the United States and Canada, but U.S. and Canadian industry has only taken small advantage of this. This is because the managers have judged statistical science in terms of some course that they took in college and which they felt was extremely dull. Only a few people have come to realize the tremendous power of the techniques involved. This is partly due to the fact that the techniques have to be adapted to the particular problem under study, and sometimes this takes a great deal of time and effort.

Another fundamental difference in Japanese quality is the way workers are treated and their relationships with their supervisors. There was a period in the early industrialization of the United States when workers were treated as adversaries. During that period the workers were for the most part undereducated. Today workers are much better educated, are much more able to operate independently of supervision, and are capable of helping supervisors in making improvements in the way their jobs are done. Both Deming and Juran say one of the least utilized resources we have in the United States is our educated work force. Nevertheless, the supervisors in the United States have not outgrown the old habits and tend not to consult the workers about changes in their jobs.

The History of Competition from Japan

The Japanese not only consult the workers about ways to improve their jobs, but they also give key workers lifetime employment. There is a class of workers in Japan who are generally operating as independent contractors and who do not have this protection, of course. However, key jobs are filled with workers with lifetime employment. The advantage of this, of course, is that if a certain part of a job is to be automated, the worker knows that he will not lose his job, and hence he cooperates with management. He knows he will be retrained, if necessary, for another job within his company. Compare that setup with the anguish that workers suffer in the United States when their company is bought out by another, and some long-time workers are fired.

The relationship of workers to managers in Japan is further strengthened by the more democratic way the managers mingle with the workers. The United States likes to think of itself as the citadel of democracy. The United States developed democracy from an earlier, more autocratic system, and vestiges of the autocratic system crept into the democracy of this country. The Japanese had democracy forced upon them after World War II, and now many of Japan's institutions are more democratic than the corresponding components in the United States. H. Ross Perot, the founder of Electronic Data Systems, has faulted the aloofness of the management of General Motors. He suggests very strongly that the managers need to get out of their ivory tower and mingle with the workers. This is good advice. It leads to a much higher level of loyalty by the workers.

Management in the United States also tends to be oriented toward the bottom line for the next quarter, or the end of the year, and does not spend enough time making long-range plans for the company. Both Deming and Juran say that 80% of a company's defects in product are directly due to failures of management, and less than 20% can be traced to the fault of the workers. These percentages appear to be the same across almost all industries.

One of the first things noted by the visitors to Japan who went there to search for the secret of Japan's success was quality circles. Quality circles are meetings organized in the work place around a small group of workers whose jobs have some elements in common. The workers discuss their jobs and how they might be improved. This has been tried in the United States with only limited success, probably because of the differences in the relationships between managers and workers. Quality circles have certainly not been the "quick fix" many hoped they would be. It appears that management style must be changed first, and then some good can be obtained with quality circles.

In short, a sweeping overhaul of the corporate culture is underway. There must be a permanent commitment at all levels, not just to top management where golden parachutes are common, but to all employees. Then all employees need to be recruited to make continuous improvements in the product or services of the organization. Permanent commitment is tentative in most companies now. Somehow this must be strengthened so that everyone in the company is convinced that it is in the personal best interests of that employee to see that the company prospers. The commitment would have to be contingent on this prosperity in most companies, but the worker would know that a prosperous company guarantees him a job. The workers would then be motivated to do the best possible job.

Too many managers in the United States spend most of their time redressing problems with their product (putting out fires), instead of worrying about how their departments should be operated, i.e., making plans for upgrading their processes. It should also be emphasized that these comments apply to service industries as well as to manufacturing and extraction industries.

3
The
Scientific
Method

THE SCIENTIFIC METHOD IS USED BY SCIENTISTS TO DISCOVER AND control variation in their fields of expertise. The commonality of this kind of study over all of the fields of science has been collected into the field known as statistical science. The basic structure of this field involves the idea of populations and samples. A random variable is a variable whose outcome changes from observation to observation (perhaps only slightly). The collection of all *possible* values that a random variable can take on is called the population.

We generally cannot obtain all of the possible values a random variable can assume, and we can never obtain them if there is a possibility of an infinite number of them. Hence, we have to rely on samples from the population to tell us about the population. If the samples are to be useful, they must be random, and in any event, every time we take a sample we are likely to obtain different values in the sample. The mean of the sample is a statistic that estimates the population mean. Since the statistic changes from sample to sample, the statistic itself has a distribution, whereas the mean of the population is a constant. This is a fundamental difference between a population's characteristics and a sample's characteristics. That is, a population and its characteristics (called parameters) are fixed, whereas a sample and its characteristics (called statistics) vary from sample to sample.

The aim of science is to explain phenomena. In order to do this, most sciences and engineering disciplines have attempted to hold all conditions constant and then measure the effect on the phenomenon by varying only one condition at a time. Many models have been developed within the various sciences by this process. In agriculture, this was not possible because the laboratory is often the open field, where many conditions vary at the same time (e.g., amount of water, sunshine, fertility, etc.), and hence, models were developed there that allow for variation of several conditions simultaneously. It was also clear in the agricultural models that there was interaction between the various conditions, and optimum results would not be obtained by varying only one condition at a time. Hence, the field of agricultural experimentation developed much more general models and achieved spectacular results. This is one of the reasons that the United States still leads the world in agriculture.

Each field of science and engineering has its own rubric, but there is much in common in each of these disciplines. The scientific method is the collection of what is common to all of the procedures of the scientific disciplines. Statistical science studies the scientific method.

Consider a contour map of the elevation of land in a given area, as shown in Figure 1. Each contour follows all of the land that is at a fixed elevation. Thus, a contour may indicate all land that is contiguous with an elevation of 500 feet and another contour may follow all of the land with a height of

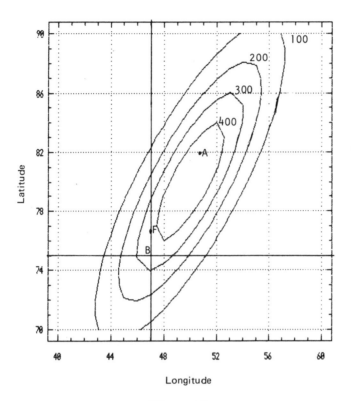

Figure 1

Contour map, showing high point at A, and one variable at a time search ending at F.

600 feet. The closer the contours are together, the steeper the slope of the land, i.e., the faster it is rising.

Now consider the situation where you do not know the map of the region you wish to explore, but you wish to find the highest point in the region, point A on the map. You can sample any point in the region you choose, but you want to find the peak with the least amount of sampling. That is, by running an experiment you can find the height at any point. Your problem is to guess which points to sample. How would you go about deciding which points to take?

This is precisely the problem that an engineer must face when he is asked to design a new product. He must probe what statisticians call the response surface to locate the best value to set as the target for his new device. Let us continue with the language of the map. Remember, the researcher does not know where the peak is located, and he/she does not know the location of the contours. The researcher can only select a latitude and longitude, run an experiment, and determine the height at that location. After doing this at several locations, she/he has to choose a point that she/he thinks is the peak.

As mentioned earlier, most scientific and engineering disciplines have been taught to hold all factors constant and vary only one factor at a time. Now return to the map and we see that if we hold the latitude constant at the value indicated and take, say, three readings at three different longitudes as indicated, we would pick location B as the highest point for the fixed latitude. Now the experimenter would hold the longitude constant at B and choose three different latitudes. The peak is now at F, and the experimenter might then set F as his target point.

Since you already know the peak is at A, you know that the researcher has been misled by his experiments. What has gone wrong here is that there has been no allowance for the interaction between latitude and longitude. The experiment was conducted in such a way that it was assumed that the effects of latitude and longitude were independent of one another. Strange as it may seem, many (and probably most) experiments are being conducted in just this way, and many products are not designed to reach their optimum at all. The reason this fallacy in approach has not been recognized by many people is because in many engineering problems the factors have tended to be essentially independent of one another. However, products have become much more complicated, and interactions are now playing a bigger and bigger role.

The solution to this problem is found in the field of agricultural experimentation, where interaction has been recognized for a long time. After all, the effects of water and sunlight obviously interact. Many experimental designs can be applied to this problem. An experimental design just tells the experimenter what configuration of points to use to obtain information to determine the optimum.

A simple experimental design that would help in this problem would be to choose three levels of latitude and three levels of longitude, and from the altitudes at these nine points, the direction of the peak is suggested. Then additional points are taken in the indicated direction and analyzed for more information on the direction to the peak. This is continued until the peak is found. In other words, you start climbing the steepest ascent. Sometimes there are ridges in the terrain and also local peaks, but a much higher peak can exist in another location. Hence this search can be very confusing, and it takes the combined experience of the experimenter and a statistical scientist to arrive at the optimum with a minimum of experimenting.

You also must realize that while we can visualize what is happening in three dimensions, as we did in the example cited, some items have many parameters to be chosen. Conceptually, these represent multidimensional searches for optima, and the experiments that must be conducted get very complicated. However, the tools are available in the subfield of statistical science known as design of experiments. The reader is referred to Chapter 13, where we will discuss this problem, and to Chapter 15, where we will discuss finding adequately trained statistical scientists.

4
Understanding Variation

WE ALL RECOGNIZE THAT INCOME VARIES FROM PERSON TO PERSON. For example, we may have the salaries of each of the individuals in our company before us. In order to make these ideas more concrete, let us suppose the annual salaries are:

$40,000
62,000
38,900
56,000
71,000

We will keep the number of salaries small in order to make the computations easy. The mean is then the sum of these salaries divided by the number of individuals involved. In this case the mean is $267,900/5, or $53,580. The mean is sometimes also called the average, but some other measures are also called averages, and hence to be specific in what is calculated we will refer to the sum of observations divided by the number as the mean. The symbol for the mean is X-bar or $\bar{X}$.

The mean tells us a great deal, but it does not tell us anything about how the salaries vary. One such measure of variability is the range, which is defined as the difference between the largest salary and the smallest salary. In this case the range is $71,000 - $38,900 = $32,100. Another such measure is the standard deviation , which has a much more complicated formula, but which may be obtained on many hand-held calculators without much difficulty. The standard deviation is the square root of the sum of squared deviations from the mean divided by the number of observations (or the number of observations minus one). The symbol for the standard deviation is sigma (σ) for a population and s for a sample. If the salaries listed are just a sample from a larger group of salaries, then the sample standard deviation, s, is equal to $13,965.39. If, on the other hand, the entire population is represented by this group of five salaries, the population standard deviation, sigma, is equal to $12,491.02.

The understanding of these three measures, range, s, and sigma, comes mainly through experience in dealing with them. However, the larger any one of these measures is, the more variable the data are. Also, if there is no variation in the data, then all of these measures are zero. They also are always positive numbers. The square of the standard deviation (either population or sample) is called the variance. You will often hear reference to the analysis of variance, which is a technique for locating sources of variation, and which will be discussed in Chapters 7 and 13.

The point that we are making is that everything varies, and we need to get used to measures of variation, just as we are used to measures of average. The range and standard deviation are in the same units of measure as the averages (and the data from which they are computed). Therefore, the interpretation of their values is always relative to these units of measure. A value of 10,000 for a standard deviation for some data may indicate small variation, whereas for other data it may indicate very large variation. For most applications, however, it is unusual to obtain observations more than three standard deviations from the mean, and it can be shown that the probability that a random variable differs from its mean by more than three standard deviations is less than 1/9. This idea is explored in Chapter 5, where we set up control limits on a process.

As we mentioned in Chapter 3, any quantity computed from the sample (range, mean, standard deviation) is called a statistic and will vary from sample to sample. On the other hand, any quantity computed for a population is a constant and is called a parameter of the population. There is no variation in the population parameters, because you have the entire group at hand, whereas with a sample you only have a part of the group at any one time. Most of the time we only have a sample at hand and we want to make inferences to some population from which the sample was taken. This process is called statistical inference.

It is clear that most people in the United States in the 1980s do not understand how variation works, how to measure variation, or how to handle problems involving variation. The increased complexity of our products and the added criticality of the target values for which they are designed make it imperative that this deficiency be overcome. We will discuss Shewhart control charts in Chapter 5. These charts provide an extremely simple way of dealing with variation.

5
Determining Whether a Process Is in Control

MANY PROCESSES MAY APPEAR TO BE UNDER CONTROL ON THE SUR-
face, but, in fact, they are not in control. Under these cir-
cumstances management tends to blame the worker for
problems when, in fact, the worker has no control over the
situation. More will be said concerning this in Chapter 10.
Control must be established in order to determine what
steps are required to improve the process.

Establishing control of processes consists of the following steps:

1. Identify the critical components of the process (those
items that lead to the most defects). Juran says that just a few
items account for 80% of the defects and many, many items
account for just a few defects. Hence it is important to identify the critical few and concentrate on them. Juran has
labeled this procedure a Pareto analysis, because the histograms showing the number of defects for each of the
causes tends to approximate a Pareto distribution. However,
there is no particular significance or usefulness to using this

name. Hence, a Pareto analysis is just a search for the critical few causes of defects. Figure 2 is a Pareto diagram for a process where five factors were identified as influencing the quality of the product. This diagram shows that humidity has the biggest effect of the five on the process, and hence humidity must be worked on first, followed by color and temperature.

2. An Ishikawa cause-and-effect diagram (also known as a fishbone diagram because of its resemblance to the skeleton of a fish) may prove useful in trying to find the sources of

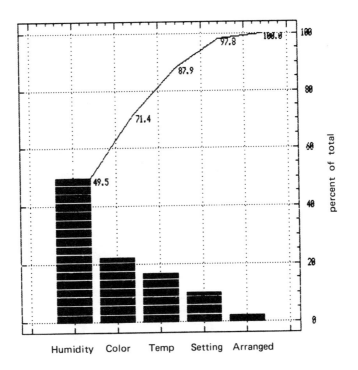

Figure 2
Pareto chart, showing relative frequency of causes of failure.

defects. An example of such a diagram is given in Figure 3. With complicated product or processes this can be a great time-saver. It helps to visualize the effects of various factors on the final product.

3. Identify measurements that may be taken to determine how the process is operating and apply one of the standard control charts to those measurements. This identification may be preceded by a Pareto analysis of the measurements.

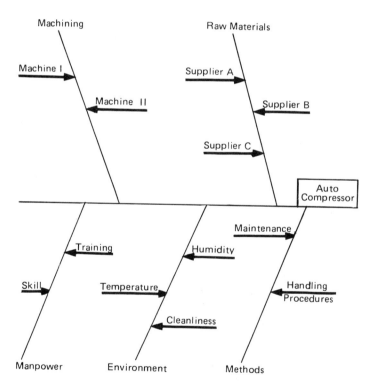

Figure 3
Ishikawa fishbone diagram or cause-and-effect diagram.

4. Make sure the process is capable of meeting the specifications on critical measurements. This is done by comparing the distribution of critical measurements to the specifications. A preliminary sample of 50 measurements is usually sufficient to establish the distribution of measurements. In general, the process is considered capable if four standard deviations either side of the mean falls within the specifications.

5. Gather some representative samples from the process to establish the control limits. The process itself should be used to establish the limits. The limits should not, as a rule, come from some specification or standard. Typically 20 or more samples are gathered before trying to establish the limits. Each sample typically contains five measurements, although any size may be used.

6. After control limits have been established, plot all of the data available on the control chart, including all of the past data which were used to establish the control limits. If any of the points used to establish the control limits fall outside the control limits, determine why they fell outside (find the assignable cause). Then recompute the limits with the out-of-control points eliminated, and repeat the whole process with new limits.

7. As new data become available, they are plotted on the control chart. Any points that fall outside the limits are examined and the cause for their being outside is determined and eliminated. There are additional triggers also used for looking for assignable causes. These will be discussed in Chapter 6.

Walter Shewhart, who was an engineer working for Bell Telephone Laboratories after World War I (in the 1920s), invented the control chart approach. His charts are simply made up so that you can take a target value and then go three standard deviations of the variable being plotted either side of the target value. These two lines are the control lines. If a point falls outside the control line, the process is declared out of control and we look for an assignable cause. Sometimes a process is continued in operation while the search for the assignable cause is underway, but most of the time the process is shut down until the cause is found and eliminated. The economics and seriousness of the situation often dictate what is done.

There are three types of Shewhart control charts. There is the system based on the mean of a variable, which also requires a chart on the range, R, or on the standard deviation, s. For this control chart (or pair of control charts) we use the following multipliers to obtain the control limits:

Sample size	A_2	A_3	D_3	D_4	B_3	B_4
2	1.880	2.659	0	3.267	0	3.267
3	1.023	1.954	0	2.575	0	2.568
4	0.729	1.628	0	2.282	0	2.266
5	0.577	1.427	0	2.115	0	2.089
6	0.483	1.287	0	2.004	0.030	1.970
7	0.419	1.182	0.076	1.924	0.118	1.882
8	0.373	1.099	0.136	1.864	0.185	1.815
9	0.337	1.032	0.184	1.816	0.239	1.761
10	0.308	0.975	0.223	1.777	0.284	1.716

For these variables charts we start with 20 to 30 samples, and for each sample we compute the mean, $\overline{X}$, and the range, R, or the sample standard deviation, s. The mean of all of the means is represented by X with two bars over it, $\overline{\overline{X}}$, and is called the grand mean. Similarly, the mean of the ranges is indicated by an R with a bar over it, i.e., $\bar{R}$, and the mean of the sample standard deviations is indicated by $\bar{s}$. The three lines on the control chart are then:

Central line	Lower control limit	Upper control limit
$\bar{\bar{X}}$	$\bar{\bar{X}} - A_2\bar{R}$	$\bar{\bar{X}} + A_2\bar{R}$
$\bar{R}$	$D_3\bar{R}$	$D_4\bar{R}$
or		
$\bar{\bar{X}}$	$\bar{\bar{X}} - A_3\bar{s}$	$\bar{\bar{X}} + A_3\bar{s}$
$\bar{s}$	$B_3\bar{s}$	$B_4\bar{s}$

Note that there is no need to compute both sets of charts. We use either the one based on R or the one based on s. The choice depends on how the variability is computed for the samples.

As an example of variable control charts consider the following data which were taken from Irving W. Burr's book *Statistical Quality Control Methods* (Marcel Dekker, Inc., New York, 1976, p. 36).

Table 1
Final Mooneys (Measure of Internal Viscosity) of
GR-S Rubber Samples of Four Successive Measurements

Average $\bar{X}$	Range R
49.750	6.0
49.375	2.0
50.250	2.0
49.875	2.5
47.250	1.5
45.000	6.0

Table 1
(Continued)

Average X̄	Range R
48.375	1.5
48.500	1.0
48.500	1.0
46.250	3.5
49.000	2.0
48.125	1.5
47.875	2.5
48.250	2.0
47.625	1.5
47.375	1.5
50.250	3.5
47.000	2.0
47.000	1.0
49.625	2.5
49.875	1.0
47.625	2.5
49.750	1.5
48.625	2.0

The grand mean for the X-bars is 48.3802, and the mean of
the R's is 2.25. Therefore, the control limits on X-bar are
50.0205 and 46.7400. Figure 4 is a control chart for the X-
bars and Figure 5 is a control chart for the R's. We notice
that samples 3, 6, 10, and 17 are out of control on the X-bar
chart and that samples 1 and 6 are out of control on the R
chart. The assignable cause for these out-of-control points
must be found. After they are found and their causes eliminated, samples 1, 3, 6, 10, and 17 are dropped from the
calculations for the control limits and new limits are found.
In this case the grand mean of the X-bars is 48.4013, and the
mean of the R's is 1.73684. Note that the five samples elim-

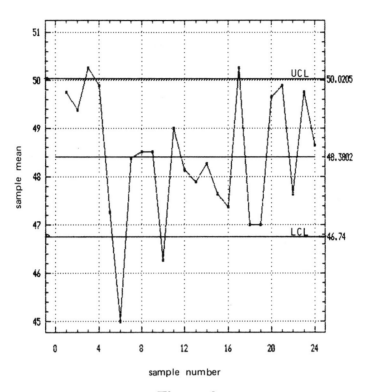

Figure 4

X̄-bar chart for final MOONEYS (measure of internal viscosity) of SR-S rubber.

inated were eliminated from both the X-bar and the R recalculation, even though not all of these points were out of control on both charts.

Figures 6 and 7 are the points replotted after the first elimination process. We see that there are samples out of control now at the new samples 2, 13, 14, 16, and 18 on the X-bar chart, although all points are in control on the R chart. Assignable causes for these points must now be sought and

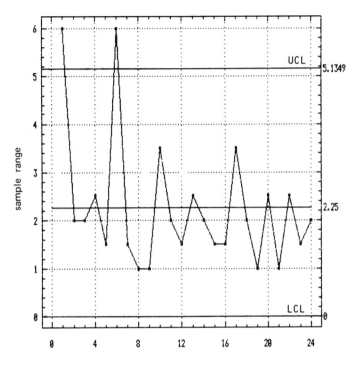

Figure 5
R chart for final MOONEYS.

the entire process repeated. After three of four iterations like this, if control is not established, additional work on the process is called for, because process control cannot be established without some change in the process.

Other control charts in common use are referred to as attribute charts since they involve counts of numbers of defects or of defectives. The distinction between defects and defectives is simply this. Think of a manufactured part, and if the part does not work it is defective. On the other hand, it

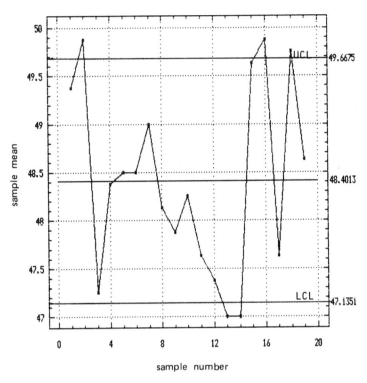

Figure 6
$\bar{X}$ chart.

can have many scratches on it which might be considered defects, but they do not keep the part from working.

The number of defectives found in a sample is usually plotted on an np chart, where p is the proportion of defectives in the sample and n is the number of items in the sample. The mean of all of the p's for 20 or 30 samples is indicated by $\bar{p}$, and the control limits are as follows:

Determining Whether a Process Is in Control

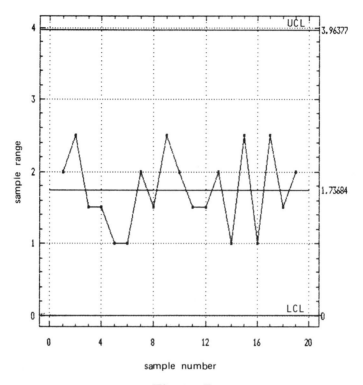

Figure 7
R chart.

Central line	Lower control limit	Upper control limit
$n\bar{p}$	$n\bar{p} - 3\sqrt{n\bar{p}(1 - \bar{p})}$	$n\bar{p} + 3\sqrt{n\bar{p}(1 - \bar{p})}$

If, instead of plotting the number of defectives, we want to plot the proportion of defectives, the limits are as follows:

Central line	Lower control limit	Upper control limit
$\bar{p}$	$\bar{p} - 3\sqrt{\dfrac{\bar{p}(1-\bar{p})}{n}}$	$\bar{p} + 3\sqrt{\dfrac{\bar{p}(1-\bar{p})}{n}}$

The number of defects found in a sample is indicated by c, and the mean of 20 or 30 such samples is indicated by $\bar{c}$. The chart for the number of defects is set up as follows:

Central line	Lower control limit	Upper control limit
$\bar{c}$	$\bar{c} - 3\sqrt{\bar{c}}$	$\bar{c} + 3\sqrt{\bar{c}}$

Data were taken from the book by Burr, *Statistical Quality Control Methods* (p. 115).

Table 2
Samples of 50 Sheets of Book Paper, 25 by 38 in., Tested for Sheet Formation Against a Carefully Selected Standard Sheet, so that if the Sheet Tested Has Poorer Formation than the Standard, It Is Called a Defective.

Day no.	Number of defectives
1	1
2	0
3	1
4	4
5	1

	Table 2 (*Continued*)
Day no.	Number of defectives
6	5
7	2
8	1
9	3
10	3
11	3
12	7
13	3
14	2
15	2
16	3
17	0
18	3
19	3
20	2
21	4
22	3
23	1
24	2
25	3
26	0
27	2
28	5
29	2
30	1

Note that we do have 30 data sets, and that we usually require 20 to 30 data sets to establish the control limits. Here
$n\bar{p} = 2.4$; Figure 8 is a plot of the data with the control limits
at 0 and 6.9. One point, sample 12, is out of control. After an
assignable cause is found for this data point, the control is
set up again with the upper control limit at 6.6, as in Figure 9.
Now all points are in control.

A number-of-defects chart or c-chart would be required

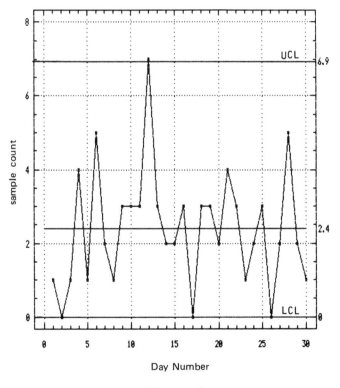

Figure 8
np chart.

for the following data taken from the book by Burr, *Statistical Quality Control Methods* (p. 131).

Table 3

Total Defects of All Kinds on Day's Productions of 3000 Electrical Switches. December, 211 working days:

Day no.	Total defects, c
1	30
2	56
3	47
4	86
5	44
6	23
7	16
8	64
9	80
10	54
11	73
12	65
13	76
14	69
15	53
16	58
17	30
18	91
19	90
20	36
21	57

The total number of defects is 1198 in these 21 samples. Therefore $\bar{c} = 57.0476$, and the lower control limit is $57.0476 - 3\sqrt{57.0476} = 34.3886$, and the upper control limit is $57.0476 + 3\sqrt{57.0476} = 79.7066$. The data are then

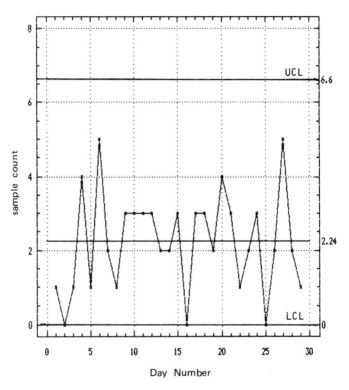

Figure 9
np chart.

plotted as in Figure 10. Days 4, 9, 18, and 19 are above the upper control limit and days 1, 6, 7, and 17 are out of control for the lower control limit. If assignable causes are found for these points and they are eliminated, the control chart given in Figure 11 is obtained. All points are now in control, but the process still needs work as the mean number of defects is high for this kind of work. Careful records must be kept on how often any particular defect occurred. These will help in determining which problems should have priority for solution.

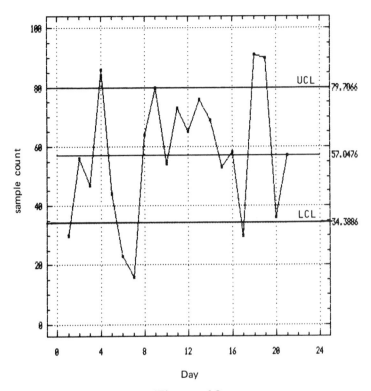

Figure 10
c chart.

One other type of control chart should be mentioned. CuSum charts are used extensively in place of X-bar charts or in place of p-charts. The average run length (ARL) is used to evaluate these charts. The ARL is the number of samples you take, on average, before the chart gives the out-of-control alarm. You want a high ARL when the process is remaining in control, and you want a low ARL when the process has moved off target. For the Shewhart control charts, the ARL is approximately 370 for the acceptable

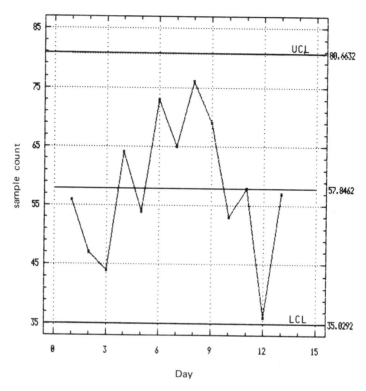

Figure 11
c chart.

quality. The Shewhart charts are not set up with any particular rejectable quality in mind. There are extensive tables of the relationships between the ARL's and the way the CuSum chart is constructed. An example of these tables is given here for replacing the np-chart of Figure 8 when the ratio of the proportion defective that should be rejected to the proportion defective acceptable is 4:

Determining Whether a Process Is in Control

	ARL (rej)								
	5.00			7.50			10.00		
ARL (acc)	m	k	h	m	k	h	m	k	h
500	0.38	0.9	2.75	0.24	0.6	2.75	0.16	0.4	2.5
250	0.32	0.8	2.25	0.21	0.6	2	0.12	0.3	2
125	0.28	0.7	1.75	0.16	0.4	1.75	0.07	0.2	1.5

The quantity m is the product of the sample size for each sample which should be taken times the proportion defective that is acceptable. The quantity charted on the CuSum chart is the sum of (number of defectives over all samples minus k). The alarm is given if this sum exceeds h. For example, if the proportion defective that is acceptable is 0.01, the proportion defective which we want to have a high assurance of rejection is 0.04, and if we want the ARL for acceptance to be 500 while the ARL for the rejectable level is to be 7.5, we see from the table that m = 0.24, k = 0.6, and h = 2.75. Our procedure then would be to solve n(0.01) = 0.24 for n to get n = 24, and then give the alarm if the sum of (number of defectives minus 0.6) is greater than or equal to 2.75, based on a sample of 24 items each time.

This procedure can be presented with the use of a V-mask to make the test for the alarm. This is often an impressive presentation in this form. The procedure can be used in place of any of the control charts above. Figure 12 gives the CuSum chart, with the V-mask showing on the right in the form of a V turned over to the left, and the last point plotted being sample 20. The exact location of the V-mask is determined by the scale used in plotting the points and certain rules about placement. This V-mask does not correspond to

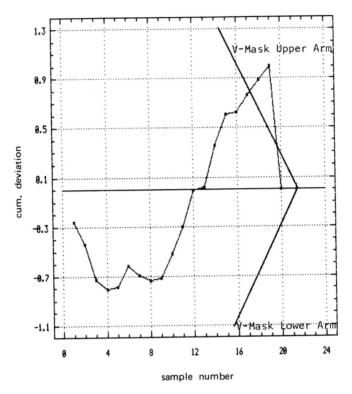

Figure 12
CuSum chart.

the data discussed in the previous paragraph. The V-mask in Figure 12 indicates that the process has gone out of control at sample 20 by showing a point or several points above the upper arm of the V. The process is also out of control if any point falls below the lower arm of the V-mask. These procedures are a great deal more sensitive to changes in the process than the Shewhart control chart, but there are also modifications which can be made to the control chart process that bring the Shewhart chart into closer competition with the CuSum chart.

6
Finding Assignable Causes

WHEN MANY POINTS ARE OUTSIDE THE CONTROL LIMITS, WE MUST LOOK carefully at the process to try to determine the causes. Sometimes the causes are internal to the way the data must have been gathered. For instance, if a product is produced by four different machines, and the output from one machine at a time is collected for a single data point, the variability that is measured is within the machine variability, and there is no allowance for machine-to-machine variability. In this case a control chart on the mean can show out of control, while the control chart on the variability is in control. If all four machines are to be plotted on the same control chart for means, then the variability from machine to machine must be added to the variability within machines when setting the control limits on the means. An obvious way to do this is to mix the output of the machines together before taking the sample. However, this may not be the best way to handle the problem. A statistical scientist should probably be consulted at this point to determine the different variabilities involved and to determine the combined variability.

Chapter 6

Another kind of problem can show up when most of the points being plotted fall too close to the center of the control chart. This also indicates lack of control, and the cause must be determined. It could be that the variability of the process has decreased in the case of a control chart on means, but then both the chart on means and variability should show this anomaly. The cure is just to recompute the control lines. In the case of a control chart on number of defects or on proportion of defectives, the cause may be more insidious. In this case some worker could be deliberately manipulating the data. The reason that the data gatherer is suspected is that the variability for these charts is directly related to the center line. You cannot have the points close to the center line for many readings, unless someone has tampered with the data in some way.

The process is also considered to be out of control when there are too many points on one side or the other of the center line. The usual rule is to call the process out of control if there are seven or more points in a row on one side of the center line or seven points in a row with each point larger (or smaller) than the preceding one (a trend is present). This can happen in several ways, including a shift in the mean and a corresponding decrease in the variability. Another way this can happen is if the machine tool used in the process has a wear factor. If the problem is wearout, then rules must be set up to reset the process whenever the product approaches a specification limit.

To find assignable causes, we must gather evidence from whatever sources are available and make inferences much like a detective. It is very helpful to enlist the aid of everyone who knows anything about the process to help in the investigation. Sometimes a worker may know of some event that seemed unimportant at the time it occurred, but that event can be the key to the problem.

7
Process in Control: Tightening Up

JUST ESTABLISHING CONTROL OVER A PROCESS IS GENERALLY NOT enough to attain real quality. There are several sources of variation for the process even after control is established. These must be identified and their impact on the process determined. The procedure is first to let the process itself establish the control limits. Second, all points outside these control limits are examined to determine the causes for lack of control. Third, these causes are eliminated from the system and new control limits are established. There is a loop at this point until all past points fall within control. Fourth, the control limits are evaluated against any specifications on the product. Note that this is not a direct comparison, because different measures of variability are usually involved. This is a point where a statistical scientist may have to be consulted as it can be confusing. Fifth, even after the process is in control and the product is meeting specifications, the process should generally be further tightened in order to get the very best quality. This tightening process will now be described.

Tightening up the process is the most important component in achieving outstanding quality. By tightening up, we
mean to bring more and more of the product closer to the
target value. This is done by analyzing the sources of variation in the process. Again, this is not unlike being a detective
in that you have to look at the product in great detail. Usually
many designed experiments are run during the detecting
process. You try to find out the effect of varying the input
parameters on the product.

If the detective work is good, after the experiments have
been analyzed you will usually find that four to six sources of
variation have caused most of the variability in the product
and many others have contributed very little. The results
may appear as in Figure 2. The idea is to choose the sources
of variation that are making the greatest contribution and
work on them first. This process has been termed a Pareto
analysis, because the configuration usually found (as in Fig.
2) can be fitted by a Pareto distribution. The terminology
here is really not informative, but it has become established
in the literature. This illustrates another one of the 80% rules
of Juran. Usually the four to six sources of greatest variation
account for approximately 80% of the variability in the
product.

The best product is obtained by a process of continually
tightening up on the control lines so that more and more of
the product is closer to the target value. This is a never-
ending process where it is practiced, but leads to extremely
reliable product. Along the way, further analysis is also made
of the original design, and that design is refined. This may
lead to a repositioning of the target value. In order to obtain
information on how the product is performing, some companies establish elaborate feedback systems from the user of
the product. One must also keep in mind that the laboratory
setting is different from the setting for the ultimate user, and

this difference may call for different target values. One caution here is not to rely on complaints from users to tell you whether your product has problems. In many settings the users merely switch to another brand when they start having difficulty with a product. The most reliable approach seems to be to follow a few product items throughout their lifetime.

Another caution is needed here. That is, it is not necessarily better to collect more data. The statistical scientist is often in the position of having to make a decision with a minimum of data and will press for additional data in order to make a reliable judgment. However, after a point which is determined by the problem being studied, having more data does not help. Some companies have collected large amounts of data on the performance of their products and then have essentially never done anything with the data. The data collected on the product should be carefully analyzed and summarized; otherwise, there is no point in collecting the data. It is wasteful to collect data that are never analyzed and summarized.

8
The Search
for Causes
of Variation

EVEN WHEN A PROCESS IS IN CONTROL, THE VARIATION IN THE OUTPUT may be more than we like. Also, to achieve the very best product we have to continue to tighten up the process. We have to hunt for the causes of variation in the product.

This search for causes of variation is different from the search for assignable causes for a process being out of control. Usually something relatively major is wrong when a process is out of control, and it can be found by a global look at the process. Much more detail is necessary when looking for the causes of variation for a process that is in control. However, this detail is easier to pinpoint because it should always be in the system, while the search for an assignable cause may be hampered by a once-only happening of an event that affected the process.

Everyone who has any acquaintance with a process needs to be consulted, and the question asked, "The output of this product varies. What do you think might be the causes of this variation?" The Ishikawa cause-and-effect diagram is often useful at this point (see Fig. 3). After several factors are identified, some experiments must be run to determine the amount of contribution from each of the factors. This is usually the point where the help of a statistical scientist is needed to design and analyze the experiments. If there is still a large component whose source is unknown, more work must be expended to search for additional factors. When the procedure for the search is complete, the amount of variation from many different factors can be listed.

9
Deming's Funnel Experiment

IN ONE OF DEMING'S FAMOUS EXPERIMENTS HE PLACES A FUNNEL over a flat surface, say a tabletop, and marks a spot on the table as a target. Then a marble is dropped through the funnel, and the place where it hits the table is noted by a mark on the table. If the funnel is centered over the target, and many drops of the marble are made, you get a pattern of dots in a circle around the target with a diameter of, say, D(1).

Now suppose we decide we want to improve the performance of the process to try to get a smaller circle around the target. We will do this by repositioning the funnel after each drop. If the marble comes to rest one inch northeast of the target, we will reset the funnel so that it is aimed at a point one inch southwest of the target, i.e., at a point at the same distance, but in the opposite direction from where the first point fell. We continue to readjust the funnel after each drop. The result is a circle with a diameter D(2), which is 41% larger than the diameter D(1) which resulted when no adjustments were made.

This is only part of the funnel experiment, but it illustrates very dramatically why the principles set forth in Chapters 6 and 7 are vital to control of a process. These principles were:(1) Find out if your process is in control. If it is in control, and the output is satisfactory, leave it alone. If the process is in control, and you wish to improve the output, you must find the sources of variation in the process and then eliminate or reduce some of them. It is impossible to improve a process merely by manipulating it. The funnel experiment illustrates that manipulating generally makes matters worse. (2) If the process is out of control, you must find the assignable causes and eliminate them before you will have satisfactory results.

An immediate application of the funnel experiment is the automatic adjustment of a process by computers. If the computer makes a measurement and immediately adjusts the process to compensate for the amount of miss of the target, the automatic feedback feature, instead of enhancing the output, actually makes the output worse than if no adjustment were made. This is not to say that all automatic adjustments will lead to worse results. You may have a process that corresponds to tool wear, for example, and if the adjustment is triggered by the movement of the average output along a trend line, then, of course, the result is worthwhile. However, whenever automatic adjustment is installed, one should look closely at what is being done and make sure that the adjustment is not making the process worse.

The funnel experiment is extended one step further. Since the above adjustment did not help, perhaps we should just take note of the last two positions of the ball on the table. Then we adjust the funnel by noting how far the ball in its last position is away from the previous position and adjust the funnel so that it is an equal distance in the opposite direction from the penultimate position of the ball. After this experiment is run many times, we see that the ball is moving further and further away from the original target. Deming says, "It is on its way to the milky way." The point to note here is that the adjustment is made without any regard to the original target point.

What does this show? Deming points out that when a new worker is trained by a worker on the job, the situation is often similar to this. That is, the new worker is often trained without any thought of the original target set for the job, and the result can be a worker who has no idea what the original target was. Another example Deming likes to cite is a Japanese auto manufacturer who used an American paint producer to make paint for its cars. Each time a new order was placed the auto maker sent in a batch of paint from the last batch with the instructions to make the new batch just like the last one. The result after several batches were made was a paint that was not at all like the original color. If you have ever had to have a fender or some other panel replaced on your car, you may have had this experience of not getting matching paint. Now you know the reason it occurs.

10
Deming's
Bead
Experiment

DURING MANY OF HIS SEMINARS DEMING CONDUCTS AN EXPERIMENT in which he uses a bowl with, say, 500 black beads and 4500 white beads. He has a paddle which can be dipped into the bowl and withdrawn with 50 beads collected in the holes of the paddle. He mixes the beads thoroughly in the bowl and then invites volunteers from the audience to come to the rostrum to dip the paddle into the bowl and withdraw a paddle full of beads. The black beads he calls defectives and the while beads nondefectives. The way this is set up there are 10% defectives being produced by the process.

Deming tells the audience that the goal is to 7% or less defectiveness. With about five volunteers, he asks each in turn to be the "worker" and withdraw the beads. He records the proportion defectiveness in each of the paddles, and he praises those workers who get less than 7% defectives and berates those workers who get more. The beads are replaced in the bowl after each draw so that the proportion defective in the bowl remains at 10%. This process is repeated about six times, and one of the workers is chosen as the "best producer for the day," because he/she got the smallest number of defectives. The worker with the most defectives is told that he should probably be considering going into another line of work.

On the average one-quarter of the workers will get 7% defective or less in one draw from the bowl, and over half of the workers will get 10% defective or more. Hence, usually most of the workers are labeled as lazy for getting 10% defective or more. Note that this is an experiment and the outcome varies as it is repeated. That is the reason for saying "usually" and "on the average." There is a very small chance that all of the workers will get 7% on a single draw.

There are two points to this experiment. (1) The worker has no control over the percent defectives generated by the paddle. (2) No amount of cajoling by management has any effect on the outcome. The outcome is strictly random.

Once a process is in control, as in this experiment, the worker cannot control the outcome. Randomness controls the outcome. The worker may want to produce good product with all sincerity, but he does not have that control. If management really wants to bring the percentage defective down to 7% from 10%, management must change the process in some way. THIS IS AN EXTREMELY IMPORTANT POINT. Management, not the worker, is at fault if a process that is in control produces more defectiveness than we want. The reader is referred back to Chapter 7 for general directions on how to handle this situation. Note that it is also necessary to find out whether the process is in control before proceeding to this point.

For workers on a process that is in control this experiment demonstrates that it is nonsense to have awards for work well done, or to pay a worker more just because she/he happened to get better results in a random draw. In fact, such awards are counterproductive. Again, we have to be sure we are dealing with a process that is in control before these conclusions can be drawn.

The idea of the preceding paragraph is extended in Deming's philosophy to include the annual review for raises. He believes that the review is counterproductive and only creates problems where none would have otherwise existed. In many companies teams of workers have been formed to solve problems. A given person may be on five teams, for example. Each of the teams is evaluated as to performance, and the individual's raise is determined by the combined evaluation of the teams upon which he has worked. This gives the worker the incentive to try to make the teams he works on the best they can be.

Differentiation between workers can also come about through promotion. It is clear that managers spend too much time agonizing over raises, and there are too many schemes in play to do a really equitable job in many businesses. There should also be a warning here that some of the schemes being sold to businesses by management firms for evaluating employees do not work well at all, and unfortunately, by the time the business realizes this, it is too late to save valuable employees.

11
Management's Responsibilities

MANAGEMENT'S FIRST RESPONSIBILITY IS TO FIND OUT WHETHER THE processes under their jurisdiction are in control. Chapter 5 explains the usual way to determine this. If the process is not in control, the reasons for lack of control must be determined and appropriate remedies applied. If the process is in control, management must realize that the individual worker cannot produce better quality without management's help. That is, the process must be changed in some way by management to obtain better quality. Just setting higher goals is counterproductive. Better product is often obtained by tightening up on the control, as described in Chapter 7.

When it comes to quality problems, management must obtain the services of one or more persons who are competent in the area of statistical science. Chapter 15 gives some sources for these people. It is folly to try to install statistical process control without expert advice. Otherwise, either the project fails or the wheel is reinvented. Either alternative is costly.

Management should work toward a goal of no defects, but not by slogans or posters. The worker cannot change those things not under his control. Deming's bead experiment is an illustration of the folly of trying to change things by exhortation.

Top management must take the lead in quality matters and must never send out signals that anything but a quality product is acceptable. For instance, giving the order to fill a commitment with less than the best quality tells the worker that management is not really interested in quality. The worker gets the idea that management's interest lies in quantity, not in quality. Top management must keep renewing their commitment to lower management and to workers in matters of quality.

The annual review of performance by workers was considered in Chapter 10, where it was pointed out that evaluating the various teams a worker is assigned to leads to more interest in the team output and decreases politicking for raises. Individuals obtain their raises through the combined rankings of the team and not ranked individually. This process can save management much time.

Middle management must stop spending time rectifying defective product and, instead, spend time on changing the processes so that defective product is decreased.

Both Deming and Juran say 80% of a business's quality problems are due to management inadequacies. That is, only management can change the processes enough to fix 80% of the quality problems. Hence, instead of blaming workers for problems, management must look at what it is doing, or not doing. The payoff is well worth the effort this takes.

Three stages of progress in the quality field are recognized. Management will gain the most from consultants if they first decide where they are in upgrading their company through statistical process control (SPC). The three stages are as follows:

Stage 1. Do you know the answers to the following questions: What is variability? How do you measure it? How do you control it? How do you reduce it? Why would you want to reduce it? If you have answers to all of these questions, then you can move on to stage 2. If not, you should ask yourself, "What problems do we have that are causing us the most grief?" The problems may be of any type, although the best results occur with problems that involve large amounts of data or conflicting data of any quantity. They do not have to be identified as SPC problems. You should then collect information on one or two of these problems to present to the SPC consultant. Once you have seen an application of the techniques to one problem, you will see that there are many other problems that will also yield to the techniques.

Stage 2. If you have moved to this stage, you undoubtedly have in-house experts. The experts may not have enough time to solve all of the problems presented to them. More likely, however, the jobs they do are limited to particular areas, such as manufacturing, when, in fact, SPC should be used throughout the company. In this stage, the consultant can be used to help define additional areas within the company where SPC can be applied.

Stage 3. If you are in this stage, your experts have served the company well, and all that needs to be done is to solve some specific technical problems, such as identifying the type of tolerance limit or mathematical function that is needed for a special problem, the best experimental design to use, and so forth. Hence, your experts probably should describe the problem and work directly with the consultant.

Chapter 11

Managers should not be putting out fires all the time. They need to have time to think about the company and how to do various jobs under their supervision more efficiently. The top executives also need to understand the techniques of SPC and how it can be applied to their company and to their own jobs. Company executives who do not follow through on this will find that their skills are obsolescent in just a few years. This applies to all types of business—not just manufacturing.

12
Translation to Service Industries

MUCH OF WHAT HAS BEEN SAID IS IN THE LANGUAGE OF THE MANUFAC-turing industries. However, almost everything is also applicable to every aspect of living. The service industries will be the next big users of these techniques, and the companies that first adopt these techniques will have a tremendous advantage over those that delay. In a few years the service industries will be in much the same position as are many of the manufacturing industries in the United States now. That is, foreign competition for service work will also develop, and the service from foreign-owned companies will be superior to that provided by American companies, unless we wake up to the challenge that faces us. We must find the sources of variability in our product and find ways to control it and reduce it.

Each separate industry will have to develop techniques for that industry in much the way that each of the scientific fields (biology, chemistry, physics, etc.) has its own rubric of the scientific method. Just remember, though, that statistical science is the depository of the commonality of these scientific methods. The challenge will be to find the particular applications of these methods to each business.

One way to proceed is to examine your business, looking for areas where there is variability. The first step is to find out whether that variability is under control, as outlined in Chapter 5. Once control is established, you need to start decreasing the variability in your critical procedures. You should ask yourself what are the causes of this variability. You may have to run some designed experiments to find out which causes are significant and which are not significant. In most instances you will find more sources of variation than you can imagine, but only a few of them will be significant, as determined by your experiments. In short, you go through all the steps outlined in previous chapters.

The techniques under discussion apply to the chief executive officer's (CEO) office as well as the sales clerk on the department store floor. In the CEO's office, perhaps the first thing to look for is where the office staff, including the CEO, spend their time. How much is spent writing letters? What classes of clients do these letters serve? Make a breakdown by categories. Is there any way some of these letters could be eliminated by anticipating a problem? What classes of visitors are there? Again, could some of these visits be eliminated by doing things differently? The idea is to free the CEO, or any other worker, from routine tasks and give him/ her time to think about the future of the business.There is much routine that can be eliminated in most offices which can give the occupant of that office time to do something more productive for the company.

It should be abundantly clear how these techniques could be applied to a dry-cleaning business or a fast-food restaurant. They can also be applied to something in as widespread use as servicing appliances where, for example, there is a technician who travels about the city repairing refrigerators. For instance, the records undoubtedly indicate what types of parts are needed most often. This then should tell you what parts to carry in the service truck. Since this changes with time, the process should be periodically updated and a record made of the number of times a particular part was not on the truck. A control chart on proportion of times parts were not available should be set up and you should determine whether or not you have control. If you have been in control and suddenly you are not, this may indicate that a new type of refrigerator is having problems.

Let me reiterate the theme of this chapter. The companies that first install these methods and learn about them will get a big jump over their competition. It takes some time to learn how to use these methods, however, and the change will not occur overnight. If you feel your company is too small to embark on this on its own, perhaps you should try to arrange for your trade association to take on the development job for your industry and make its results available to all members of the association. Two books mentioned in the list for further reading in Chapter 16 are devoted to the service industries. Neither of these books is definitive, but they do indicate the scope of the future impact of the SPC techniques on the service industries. These books are authored by Rosander (17) and by Latzko (22).

13
Designing
for Quality

AS WE MENTIONED PREVIOUSLY, IT HAS BEEN THE PRACTICE IN MANY industries in the past for design engineers to isolate themselves from manufacturing engineers. Manufacturing engineers also have ignored the problems of field engineers who either work for the customer or are sent to where the customer is using a product in order to resolve problems with the product. Statistical process control requires that all of the steps in the process be considered at the design phase of the product. Genichi Taguchi, from Japan, is credited with the idea of off-line process control, although several American and British statistical scientists had developed techniques which are not unlike those proposed by Taguchi.

The major expense in developing a new product line is in designing the manufacturing process, not in designing the product. The problem in the manufacturing process is that a product can be affected by any number of factors that are under the control of the manufacturing engineer. But where should each of these parameters be set in order to attain the highest quality with the least cost? This is a difficult problem for complicated new product lines. The interaction between the various factors that can be controlled must be taken into account, and as we mentioned in Chapter 8, extracting information from experiments of this kind is where the expertise of a statistical scientist is needed.

The process of designing experiments to determine the optimum is the subject of intense study. The statistical scientists who worked with agricultural problems developed a whole series of designs for various situations, and these have been extended by other statistical scientists to apply to many situations besides agriculture, although the original agricultural terminology often still prevails. Most of the techniques fall under the general heading of analysis of variance (ANOVA), although evolutionary operations (EVOP), (which was developed by G. E. P. Box, a native Britain, at the University of Wisconsin), and some other techniques are also used. The Taguchi method was developed independently of the other methods and is the subject of intense investigation and some controversy in the late 1980s.

The terminology includes: one-way classification, two-way classification, n-way classification, factorial, fractional factorial, randomized block, split plot, Latin square, incomplete block, fixed effects, random effects, mixed effects, multiple comparisons, multiple-range tests, and so forth. Knowing some of these terms (even without knowing the meaning) may help a manager in screening for statistical scientists to work on the problems being discussed here.

The growth in computer technology has greatly aided the design of both the product and the manufacturing processes. Many of the techniques mentioned in the preceding paragraph are now integrated into computer-aided design packages. The problem with implementing most of them is that the engineers do not have enough background in statistical science to start off cold with an entirely new product using these statistical designs.However, this is easily overcome by having a statistical scientist periodically check to see whether the technique the engineer is using is the best for the problem he has. In the past engineers tended to concentrate on just one type of product, and as long as that is the case, the setup can be much the same when the product changes only slightly. However, when an entirely new product is being prepared, it is prudent to have a statistical scientist check to be sure the right "medicine" is being given.

Chapter 16 includes several references to books on designing experiments including Anderson and McLean (12), Barker (16), Wheeler (26), and Khuri and Cornell (30).

14
Quality Control
and
Quality Assurance

ONCE CONTROL OF A PROCESS HAS BEEN ACHIEVED AND THERE HAVE been no problems over a period of time, and if further improvement is neither necessary nor desirable, then it should be clear that the control can be dropped and the effort directed to some other problem area. In other words, there should be periodic checks on all control processes, and wherever it is reasonable, controls should be dropped. In many companies where this is not done, many of the controls were found to be unnecessary when an audit was made.

As originally conceived, quality control was the procedure of setting up control charts as a product is being manufactured. Quality assurance involved checking a product after manufacture to be sure it met the needs of a customer. Quality assurance involved a check on the entire design and manufacturing process, including the distribution of the item to the ultimate consumer. Quality control was analogous to the bookkeeping function, and quality assurance was analogous to the auditing function.

The past tense was used in the previous paragraph because this original distinction has become so blurred that a person visiting a new company can no longer tell whether "the quality assurance department" actually is an auditing group or not. Apparently many businesses liked the term quality assurance better than quality control and named their quality department quality assurance, when, in fact, it was set up principally as a quality control department. This led to the unfortunate situation where many businesses have failed to realize the importance of the audit function for their products. The nomenclature misled them into thinking they had something they did not have.

Quality is not just the job of the quality department. Everyone within the business must be dedicated to the idea of producing a quality product. It is extremely important that top management insist on quality, and top management must be very careful not to send out any signals that indicate any other approach. For example, when management insists that a product that is known to be inferior be shipped, the workers become cynical about management's commitment to quality.

In order to emphasize this, the organization for the business should show that the quality function has direct access to the top management. In many businesses the quality control department and the quality assurance department report to a vice president with access to the chief executive officer. This is not a bad arrangement, but a better one is to have quality control report to the head of manufacturing and quality assurance report to a separate top officer. The reason is that quality assurance's function is to audit quality control as well as the design function, and so forth. Some problems could be hidden within an organization by having both the bookkeeping-type operation and the auditing function report to the same officer.

Quality Control and Quality Assurance

The quality control department should act as a coordinator for all tasks of a quality nature performed by all the other departments in a business. The marketing department's research of a competitive product should be coordinated by the quality control department, as well as the design department's work on quality. For instance, it is often possible to build into designs certain extra features that greatly facilitate testing for quality. The role of quality control is as an advisor, not as the commander of these operations. Many service calls from consumers involving warranties are due to misuse or neglect. In many instances it is easier to foolproof a product than it is to get the consumer to care for it.

15
Sources of Consultants and Employees

ALL DISCIPLINES HAVE PROBLEMS WITH UNQUALIFIED INDIVIDUALS taking on the role of experts in the field. The public is probably more aware of this in the medical profession because of the publicity that usually attends the unmasking of the "fake doctor." The problem is worse in the quality area by at least an order of magnitude, because there have not been stringent requirements set forth for quality personnel, especially regarding keeping up with current thinking in the area. There is also no mechanism for unmasking the fakes and the incompetents, as there is in the medical profession. As we have pointed out throughout this book, the field has changed drastically in the last 20 years. Unfortunately, many people in the area, including many recent graduates of our colleges, are still operating under the standards of the 1950s and 1960s. This is because many colleges have not upgraded their curricula to reflect the latest thinking. The problem is especially bad in the business schools and most engineering schools.

There are people operating in the quality area who claim

to have Ph.D. degrees that were given to them by correspondence schools whose only requirement is that they write a thesis on their life experiences. A far greater number of workers in the field do not understand most of the aspects of the new era in quality control and have not made adequate attempts to upgrade their knowledge. Some of these people are managers for quality for large corporations and typically after retirement have set up consulting businesses in quality. In short, the message is to be very careful whose advice you follow. Make sure the individual understands all aspects of the new type of quality as outlined in this book. If the person does know the details of the new quality and has a good background in statistical science, that person will be very valuable in designing the experiments to tighten up on control. This is the area where the biggest impact can be made in the long run.

There are many programs in statistics, as opposed to statistical science, at universities around the country. Most of these programs are directed at a particular application, such as business or sociology. When programs have this single orientation, generally the graduates do not get a broad enough picture of statistical science to do an adequate job of implementing statistical process control (SPC). However, the graduates of these programs are very useful in helping to implement SPC programs under the direction of people with a broader understanding of the field of statistical science.

We have tried to make it clear that in order to install a good program of SPC, the help of a statistical scientist will be required. Only a person with a great deal of training in dealing with data and who understands the vagaries of statistical analyses can do a reliable job in untangling the relationships between the parameters of most processes. Training in the use of statistical packages, such as SAS.

SPSSx, Statgraphics, and so forth, can be helpful, but there will be certain problems which require the advice of an expert who has gone beyond the point of knowing how to read computer output. Also, it would be a bad mistake to rely on management trainees, industrial engineers, quality control engineers, computer scientists, or social scientists for this expertise. For, although they may have some experience in this area, it will take a person with wide exposure to the thousands and thousands of techniques that have been developed to bring a business up to its peak performance. In short, a person with a doctorate or the equivalent (gained through experience) in statistical science is needed for best results. Deming says that if a company has the opportunity to hire a statistical scientist, it should surely do so. The company should then give the person hired access to the overall problems of the company by working with several teams of individuals who are dealing with quality problems.

A survey was conducted in April 1981 by a committee of the Conference Board of Associated Research Councils. The survey was an attempt to rate (by peer review) doctoral programs in the various sciences in the United States. Various people around the country were contacted and asked to rate certain schools other than their own. The results were published in *The Chronicle of Higher Education* on September 29, 1982. In the field of statistical science some schools were not included in the list which had more reason to be there than some which were included. However, the following is a list of schools included and their rankings.

	Rank
American U.	
Boston U.	
U. of California, Berkeley	1

	Rank
U. of California, Los Angeles	21
U. of California, Riverside	
Carnegie-Mellon U.	37
Case-Western Reserve U.	
U. of Chicago	33
Colorado St. U.	30.5
Columbia U.	16
U. of Connecticut	
Cornell U.	38
U. of Delaware	
U. of Florida	35.5
Florida St. U.	6
George Washington U.	
U. of Georgia	11.5
Harvard U.	18.5
U. of Illinois	39
Indiana U.	
U. of Iowa	26.5
Iowa St. U.	4
Johns Hopkins U.	21
Kansas St. U.	30.5
U. of Kentucky	18.5
U. of Maryland	
U. of Michigan	23.5
Michigan St. U.	30.5
U. of Minnesota	30.5
U. of Missouri, Columbia	34
U. of Missouri, Rolla	
U. of New Mexico	
U. of North Carolina (Biostat)	5
U. of North Carolina (Stat)	11.5
North Carolina St. U.	13.5
Ohio St. U.	16
Oklahoma St. U.	13.5
Oregon St. U.	16

Sources of Consultants and Employees

	Rank
U. of Pennsylvania	
Pennsylvania St. U.	21
U. of Pittsburgh	
Princeton U.	26.5
Purdue U.	8.5
U. of Rochester	26.5
Rutgers U.	35.5
Southern Methodist U.	23.5
U. of South Florida	
Stanford U.	3
State U. of New York, Buffalo	
State U. of New York, Stony Brook	
Temple U.	
Texas A&M U.	8.5
U. of Texas Health Science Center, Houston	
Virginia Commonwealth U.	
Virginia Poly. Inst. & St. U.	7
U. of Washington	10
U. of Wisconsin	2
U. of Wyoming	
Yale U.	26.5

The rankings indicated are for the number of doctorates in statistical science produced at the school from 1976 to 1980. Thus, the University of California at Berkeley produced the most doctorates in statistical science. Only 39 schools were ranked. Information on doctorates produced was lacking on the others, and in most instances this indicates that the school was not actually producing statistical scientists. In general, the number of doctorates produced in statistical science was small. There are some surprising omissions from this list, but it is clear that only a few schools in the United States have viable doctorates in this science which is so vital to the future of the United States.

Number of doctorates produced should not be used as a criterion for selection for any purpose other than availability. In any event, the number of people with bachelors and masters degrees in statistical science closely parallels the doctorate production. The people with these degrees (bachelors and masters) can be invaluable in carrying out the details of the analyses which are required. They should either be experienced (five years or more) or they should do this work under the supervision of a person with more background. This indicates a great shortage of people who can do the work needed to install effective SPC. Also, not everyone who graduates from these programs has the interest or training needed to do a good job here. There are other schools which might be considered, but generally the above is a core list of schools. In short, being on the list is neither necessary nor sufficient to ensure that the school's graduates would make good quality personnel, but it is an indicator.

16
Notes for
Further Reading

THE FOLLOWING BOOKS AND ARTICLES WERE CHOSEN AS APPROPRIATE to give additional information on the quality revolution. Books written before 1980 generally do not recognize the quality revolution which is underway. However, they do give the techniques that are being used in the revolution.

1. *The Keys To Excellence,* 2nd ed., by Nancy R. Mann, Prestwich Books, Los Angeles, 1987.

This small book tells the story of W. Edwards Deming's visits to Japan and his consulting with American corporations. This is a nontechnical history written on a personal level.

2. *Quality, Productivity, and Competitive Position,* by W. Edwards Deming, Massachusetts Institute of Technology Press, Cambridge, 1982.

This is one of Dr. Deming's books that describes his philosophy for quality.

3. Quality, Special Report, *Business Week,* June 8, 1987.

This issue of the periodical has several articles on the revolution taking place in quality, including "The Push for Quality," "A Tale of Two Industries," "Making It Right," and so forth.

4. *Quality Planning and Analysis*, by J. M. Juran and Frank M. Gryna, Jr., McGraw-Hill Book Co., New York, 1980.

This book concentrates on management and organization problems associated with the quality functions. Interspersed in the management discussions are the standard quality control techniques.

5. *Quality Control Handbook*, 3rd ed., edited by J. M. Juran; Frank M. Gryna, Jr., and R. S. Bingham, Jr., associate editors, McGraw-Hill Book Co., New York, 1979.

This is a classic compilation of methods and recommended organization for quality. It is a massive book of 49 chapters, some chapters containing 70 pages.

6. *Elementary Statistical Quality Control*, by Irving W. Burr, Marcel Dekker, Inc., New York, 1979.

This is a book for beginners and for high-school students. It introduces the Shewhart control charts and acceptance sampling at an elementary level.

7. *Statistical Quality Control Methods*, by Irving W. Burr, Marcel Dekker, Inc., New York, 1976.

This is a technical book written at the intermediate level. It covers the Shewhart control charts very thoroughly, as well as acceptance sampling, including a detailed description of MIL-STD 105D, or the American, British, and Canadian (ABC) sampling scheme.

8. *Acceptance Sampling in Quality Control*, by Edward G. Shilling, Marcel Dekker, Inc., New York, 1982.

This is a technical book written for the professional which views quality control techniques from the acceptance sampling point of view.

9. *Quality Control and Industrial Statistics,* 4th ed., by Acheson J. Duncan, Richard D. Irwin, Inc., Homewood, IL, 1974.

This is a classic reference book for the standard techniques of quality control. It contains over 1000 pages.

10. *Statistical Quality Control,* 4th ed., by Eugene L. Grant and Richard S. Leavenworth, McGraw-Hill Book Co., New York, 1980.

This is a classic text which has been used by many engineering schools to teach quality control.

11. *Encyclopedia of Statistical Sciences,* edited by Samuel Kotz and Norman L. Johnson; Campbell B. Read, associate editor, John Wiley & Sons, Inc., 1982 through 1987 and later.

At the time this is written, in June 1988, eight volumes of this work have been published, with more announced. This is a definitive reference work for the areas of statistical science.

12. *Design of Experiments, a Realistic Approach,* by Virgil L. Anderson and Robert A. McLean, Marcel Dekker, Inc., New York, 1974.

This gives an easily implemented (relative to other texts) approach to designing and analyzing experiments, including factorials, Latin squares, split plots, fractional factorials, and response surface exploration.

13. *Designing for Minimal Maintenance Expense: The Practical Application of Reliability and Maintainability,* by Marvin A. Moss, Marcel Dekker, Inc., New York, 1985.

Cost of maintenance has become an important component in total cost, especially for complex systems. It has therefore become more important than ever before to build systems that are reliable, easily maintained, and available for use over extended periods of time. This book tells how this can be done.

14. *Quality Control for Profit*, 2nd ed., revised and expanded, by Ronald H. Lester, Norbert L. Enrick, and Harry E. Mottley, Jr., Marcel Dekker, Inc., New York, 1987.

This book is directed at the inspection aspects of a quality program. It contains many examples and practical ideas on the implementation of a quality program. Costs of quality are emphasized. The use of designed experiments is advocated, but details on how to do this are not included in this book. The beginner will find the sections on statistics and reliability interesting, but practical application will require additional research elsewhere.

15. *QCPAC: Statistical Quality Control on the IBM PC*, by Steven M. Zimmerman and Leo M. Conrad, Marcel Dekker, Inc., New York, 1987.

This is a series of programs for the IBM-PC for determining sample size, and for plotting histograms, operating-characteristic curves, and control charts.

16. *Quality by Experimental Design*, by Thomas B. Barker, Marcel Dekker, Inc., New York, 1987.

This is a book on experimental design at a lower level than the Anderson and McLean book. People who are not trained in statistics will probably find Barker more easily read than Anderson and McLean. On the other hand, it relies on intuitive and heuristic reasoning and attempts to simplify complicated concepts without completely succeeding. Consequently, it raises the ire of some trained in mathematics and the logical development of topics. It is useful as an overview where details are ignored.

17. *Applications of Quality Control in the Service Industry,* by A. C. Rosander, Marcel Dekker, Inc., New York, 1987.

This book describes how statistical process control applies to the service industries as well as the manufacturing industries. Many examples are given of actual service industry applications. It does not give a road map for new applications, however.

18. *Integrated Product Testing and Evaluating: A Systems Approach to Improve Reliability and Quality,* revised ed., by Harold L. Gilmore and Herbert C. Schwartz, Marcel Dekker, Inc., New York, Inc., 1987.

Organizational planning, data reporting systems, control of the environment, and statistical methods are all combined in an overall systems approach. This book gives concepts and details on product development from design to customer use. Case histories are also included. A concise overview is given of the shift away from failure fixing.

19. *Quality Management Handbook,* edited by Loren Walsh, Ralph Wurster, and Raymond J. Kimber, Marcel Dekker, Inc., New York, 1987

This book is highly recommended by Phillip Crosby, one of the latter-day "gurus" of quality. It stresses the importance of management's involvement in quality. The authors compare their book to an encyclopedia; however, nearly one-quarter of the chapters contain no references.

20. *Statistical Process Control: A Guide for Implementation,* by Roger W. Berger and Thomas Hart, Marcel Dekker, Inc., New York, 1987.

This book is for managers and executives who lack the

basic ideas of statistical process control (SPC). These ideas are covered in a simple and straightforward manner. The book explains how to start an SPC program. It also explains such statistical concepts as range, distribution, and assignable causes. A glossary of terms is included.

21. *Quality Circles: Selected Readings*, edited by Roger W. Berger and David L. Shores, Marcel Dekker, Inc., New York, 1987.

A quality circle is a small group of workers at the shop level who meet to solve quality problems. This book gives a comprehensive overview of the philosophy of quality circles and discusses how statistical process control (SPC) and quality circles have succeeded for the Japanese work force. Case studies from General Motors and General Foods are included.

22. *Quality and Productivity for Bankers and Financial Managers*, by William J. Latzko, ASQC Quality Press, Milwaukee, WI, 1986.

This book discusses how to introduce quality into financial transactions and gives meaning to the relationships among quality, productivity, and cost. Control of clerical quality is specifically covered, and the differences and similarity to manufacturing quality are discussed. This is one of the few efforts to describe quality concepts in a service industry setting. See also the book by Rosander in this compilation.

23. *Poor-Quality Cost*, by H. James Harrington, ASQC Quality Press, Milwaukee, WI, 1987.

Quality products provide greater return on investment and tend to increase a company's product share. The costs of poor quality are, in most cases, such that it is less expensive to provide high-quality products than to do otherwise. When quality problems are solved, cost and schedule problems are greatly reduced. This is a book for managers and financial workers who need to know the costs in dollars and cents for less than the best quality. The cost the customer incurs as a result of poor quality is also considered.

24. *Human Resource Management,* edited by Jill P. Kern, John J. Riley and Louis N. Jones, Marcel Dekker, Inc., New York, 1987.

This book provides the reader with an overview of some of the important tools needed in dealing with people in the new quality environment. Included are chapters on preparing an organization for change, human factors engineering, quality circles, management training, strategic planning, organizational structures, and so forth.

25. *The Good and the Bad News About Quality,* by Edward M. Schrock and Henry L. Lefevre, Marcel Dekker, Inc., New York, 1988.

This book covers many of the same topics included in the book you just read. It assumes the reader understands variation and consequently will be interested in topics such as process capability, regression, designing experiments, and so forth.

26. *Tables of Screening Designs,* by Donald J. Wheeler, Statistical Process controls, Inc., 7026 Shadyland Drive, Knoxville, TN 37919, 1987.

A set of designs for experiments is given so that the reader can conduct a designed experiment by relating the problem at hand to the setup given in the tables. The designs are for two and three levels of the factors to be analyzed.

27. *Guide to Quality Control,* by Kaoru Ishikawa, Asian Productivity Organization, and JUSE Press, Tokyo, 1982. Also Kraus International Publications, White Plains, NY.

Chapter 16

This is an elementary introduction to quality control procedures with emphasis on the Ishikawa fishbone diagrams. This book has been used in many industries to introduce quality ideas. It has a unique feature in that the last chapter includes practical problems keyed to the earlier chapters of the book.

28. The Productivity Paradox, Special Report, *Business Week*, June 6, 1988.

Several articles in this special report deal with the questions of automation and why automation has not paid off well in the United States. Old-fashioned methods of cost accounting are keeping some businesses from taking full advantage of automation. The same lame excuses are being given for the lack of payoff as are given for not competing by building better-quality product.

29. Product Quality: An Important Strategic Weapon, by David A. Garvin, *Business Horizons*, March-April 1984, pp. 40–43.

A good working definition of quality is difficult to obtain. This article examines several definitions and points out the shortcomings of many commonly used definitions.

30. *Response Surfaces: Design and Analysis*, by Andre Khuri and John A. Cornell, Marcel Dekker, Inc., New York, 1987.

This book tells a reader how to climb a mountain when the reader only knows roughly where the mountain is located. This is the problem faced when trying to design a product or service in an optimal way.

31. *Statistical Quality Control for Manufacturing Managers*, by William S. Messina, John Wiley & Sons, Inc., New York, 1987.

This book includes material on the Pareto diagrams, and Ishikawa cause and effect diagrams, and also more about statistical process control as opposed to quality control, which is found in older books.

32. *Modern Methods for Quality Control and Improvement,* by Harrison M. Wadsworth, Kenneth S. Stephens, and A. Blanton Godfrey, John Wiley & Sons, Inc., New York, 1986.

This book gives a very comprehensive review of the steps involved in statistical process control. It is not intended for the beginner, however.

33. Guide for Quality Control Charts, Control Chart Method of Analyzing Data, Control Chart Method of Controlling Quality During Production, *American National Standard,* ANSI Z1.1-1986, Z1.2-1985 & Z1.3-1985, ASQC STANDARD B 1-1985, B2-1985 &B3-1985, American Society for Quality Control, Milwaukee, WI

The title and publisher tell the content of this publication.

34. Definitions, Symbols, Formulas, and Tables for Control Charts, *American National Standard, American National Standard,* ANSI/ASQC A1-1987, American Society for Quality Control, Milwaukee, WI.

The title and publisher tell the content of this publication.

35. *Practical Statistical Sampling for Auditors,* by Arthur J. Wilburn, Marcel Dekker, Inc., New York., 1984.

This book tells how auditors use statistical methods in doing their work.

Index

A

Adversaries, 12
Agriculture, 20, 23, 94
Alarm, 51
American Society for Quality Control, 121
Anderson, 115
ANOVA, 94
ARL, 49, 51
Assignable cause, 5, 36, 37, 39, 40, 48, 55, 56
Attribute charts, 41
Audit, 99, 100, 101
Automation, 13
Average, 27, 49
Awards, 76

B

Barker, 116

Bead experiment, 75, 76, 77
Bell Telephone Laboratories, 36
Berger, 117, 118
Best producer, 75
Bingham, 114
Box, 94
Burr, 38, 44, 45, 47, 114
Business Horizons, 120
Business Week, 113, 120

C

c-bar, 44
c-chart, 44, 46, 47, 48, 49
Capability, 36
Capital from savings, 6
Cause-and-effect diagram, 34, 35, 65
CEO, 88

Index